Problemoids
Math Mentor

LEVEL 5

Bill McCandliss
Albert Watson

Royal Fireworks Press
Unionville, New York

Acknowledgments

Since before recorded history, Man has posed and solved mathematical problems. The authors of *Problemoid* are indebted to a long tradition of men and women—some famous, many unknown.

In particular, we wish to acknowledge and pay tribute to George Polya, whose identification of problem-solving strategies influenced the Hints section of *Problemoids*.

Bibliography

Bloom, B. S., editor.
Taxonomy of Educational Objectives, Handbook I:
Cognitive Domain.
New York: David McKay, 1956.

Gallagher, James J.
Teaching the Gifted Child.
Boston, MA: Allyn and Bacon, 1975.

Polya, George.
How To Solve It: A New Aspect of Mathematical Method, Second Edition.
Princeton, NJ: Princeton University Press, 1957.

Copyright © 2001, Royal Fireworks Publishing Co., Inc.
All Rights Reserved. Copying of any page in this manual is prohibited.

Originally published by Trillium Press, Inc.

Royal Fireworks Press
First Ave, PO Box 399
Unionville NY 10988
845 726 4444
Fax 845 726 3824
Email: mail@rfwp.com
Website: rfwp.com

ISBN 13: 978-0-89824-044-3
ISBN 10: 0-89824-044-1

Printed and bound in the United States of America by American citizens using recycled, acid-free paper, vegetable-based inks, and environmentally-friendly cover coatings at the
Royal Fireworks Printing Company of Unionville, New York.

Contents

About the Cover 4
Implementation 5
Background 7
Problemoids & Answers to Cover Questions 8
Appendicies 9-13
Problems   1. Yin & Yang
          2. Inflation
          3. Dribblers
          4. A New Game Plan
          5. What's the Difference?
          6. Where There's a Will, There's a Way
          7. Bit Buy Byte
          8. Confronting the Clock
          9. Candydextrous
         10. Aw Sum Gauss
         11. The Muse Family
         12. A Ball with the Bearings
         13. Sexagenarian
         14. The Toothpick Fairy
         15. Lots of Dots
         16. A Liter Expedition
         17. Splitting the Spoils
         18. Sweet Enigma
         19. Crack the Cube
         20. Fifty Point Question
         21. A Little More
         22. Check 'Er Balance
         23. Caveat Emptor
         24. M & M's
         25. Hoppin' to Win
         26. The Case of the Quick Bookcases
         27. Hex Symbol
         28. Pause for Claws
         29. The Wandering Bishop
         30. Computing the Odds
         31. Distributing the Wealth
         32. Friendly Finish
         33. Getting Even
         34. An Almost Perfect Place
         35. Chew On This One
         36. A Walking Conundrum
         37. Hundreds Censored!
         38. Try Your Luck
         39. Quick Candlestick
         40. Partly Popular
         41. An Odd Gift
         42. Fair Ball?
         43. Creating Cranapple
         44. Relatively Unknown
         45. Name Dropping
         46. Parsimonious Purchaser
         47. Installation Free
         48. Quintillion Plus
         49. Three-Quarter Dime Jingle
         50. No Heads is Better Than One

About the Cover:
Pascal and His Arithmetic Triangle

Blaise Pascal (1623-1662) was born in Clermont-Ferrand, France. He was an exceptionally bright child, inclined toward mathematics, but at that time his father thought that studies in this area might harm his son's health, so Pascal had to be content with the study of languages and what mathematics he could discover for himself. Fortunately for us, Pascal was a very curious young man intrigued by mathematics, especially geometry. By the time he was twelve years old, Pascal deduced several important properties of triangles even though he never had seen any books on the subject. When Pascal's father realized his son had extraordinary talent in mathematics, he relented and encouraged his son's pursuit of mathematical knowledge.

The group that eventually became the French Academy of Sciences recognized Pascal's genius by inviting him to participate in its weekly discussions when he was just fourteen years old. Pascal devoted his early years to mathematical and scientific experimentation. When he was 19 he invented the world's first digital calculator to help his father who was a tax administrator. This mechanical device was a collection of wheels and gears which performed addition by carrying automatically. As recently as 1960 mechanical addition-subtraction devices based on Pascal's design sold for less than a dollar in the United States.

Pascal, during his short life, made important contributions to several branches of mathematics. In 1640 he wrote an essay on what he called "The Mystical Hexagram." This work, a study of the connections between conic sections and projective geometry, won Pascal wide acclaim and is still studied by mathematicians today. One part of this work can be stated simply:

Draw a circle; pick six points on the circumference and label them clockwise A, B, C, D, E, F. Form a hexagon by drawing line segments AB, BC, CD, DE, EF, and FA. Extend segments AB and DE until they meet at a point. Likewise, extend BC and EF and extend CD and FA. Call the three points of intersection X, Y, and Z. Pascal proved that points X, Y, and Z must lie on a straight line.

During his lifetime Pascal exchanged ideas with many prominent mathematicians including Gerard Desargues, Rene Descartes, Marin Mersenne, and Pierre Fermat. His communication with Fermat through letters led to the creation of the theory of probability.

One of Pascal's most famous contributions was his discovery of many fascinating properties of what is now called "Pascal's Triangle." The illustration on the cover of *Problemoids* depicts part of Pascal's Triangle. The triangle has 1's at the top (or apex) and at the ends of each row. Every other number in the triangle is the sum of the two nearest numbers in the row above it:

```
            1
          1   1
        1   2   1
      1   3   3   1
    1   4   6   4   1
```

Pascal called this triangle the Arithmetic Triangle. He did not design the triangle—there is evidence of its use by Omar Khayyam in the twelfth century—but it bears his name because he discovered many of its properties.

The cover of *Problemoids* illustrates one of the properties. Using two colors, one for the even numbers and one for the odd numbers, interesting triangular patterns appear. Notice the white triangles running through the center of Pascal's Triangle. The top "triangle" has one white dot, the second has six white dots, and the third has twenty-eight white dots. Two of these numbers, 6 and 28, are "perfect numbers;" that is, each is the sum of its divisors (excluding itself): $6 = 1 + 2 + 3$ and $28 = 1 + 2 + 4 + 7 + 14$. You might want to draw an even larger version of Pascal's Triangle than the one on the cover.

The rows and diagonals of Pascal's Triangle have surprising and useful properties some of which can help you solve several problems in the *Problemoids* program. For example, suppose you add the numbers in any diagonal down to some point on that diagonal. There is an easy way to find the sum without performing the actual addition. Add the numbers on the third diagonal down to 45. Notice that $1 + 3 + 6 + 10 + 15 + 21 + 28 + 36 + 45 = 165$ and 165 is the number on the fourth diagonal just below 45. Try several other diagonal sums. You'll find the same thing happens every time.

You may discover a few other properties as you solve the problems below:
(1) What is the sum of the numbers in the tenth row of Pascal's Triangle? (It is customary to call the top row of Pascal's Triangle the 0th row. The tenth row begins: 1, 10, 45, ...)
(2) How many numbers in the thirteenth row are divisible by 13?
(3) How many odd numbers are there in the thirty-first row?

The answers to these problems appear at the end of the book.
NO PEEKING!

IMPLEMENTATION

The *Problemoids* program does not demand significant teacher involvement, although teachers may choose to participate quite actively. In either case the following recommendations will suffice for management of the programs.

(1) ASSIGN THE PROBLEMS IN SEQUENCE.

Until the teacher becomes familiar with the problems, hints, answers, and solutions in *Problemoids*, we recommend assigning problems according to the order in which they appear in the student problem book. Generally, the position of the problem in sequence reflects its level of difficulty; that is, less difficult problems appear earlier and more difficult problems appear later in the sequence. Also, students' understanding of a problem appearing later in the sequence may be contingent on their familiarity with an earlier problem.

(2) DIRECT STUDENTS TO MAINTAIN A RECORD OF THEIR WORK.

Students should maintain a record of their work toward solutions of problems. Space for work on each problem has been provided in the student book to assist students in organizing and keeping up with their solutions. The solution of a problem may be attempted over several sittings. For this reason, students need one fixed workspace to help them organize the solution and recall what they attempted in previous work.

If students do not work in their problem books, they should maintain their work in individual notebooks. A record which shows how individual problems were solved and how various strategies were employed may be beneficial for solving future problems.

(3) DIRECT STUDENTS TO THE HINTS SECTION OF THE BOOK AND ENCOURAGE MODERATE USE OF THE HINTS.

If students feel blocked in their solution of a problem, they should be encouraged to use one or more hints. Hints to a given problem are not grouped together, but appear separately to encourage using hints one at a time. For your convenience, the Teacher's Manual has the hints grouped together. Students probably will obtain the most benefit from a hint by allowing themselves time for reflection instead of reading through all of the hints to a problem quickly. Since this factor is difficult to control, the hints have been designed to provide instructional value in either case.

(4) ALLOW STUDENTS AMPLE TIME TO THINK ABOUT AND SOLVE EACH PROBLEM.

The authors suggest assigning one or two problems at the beginning of a week. Since many of the problems are quite challenging, students usually will not be able to solve an individual problem during one class period. Students will need to spend time understanding a problem, investigating possible approaches to its solution, reflecting upon how to utilize hints, debugging unsuccessful solutions, and satisfying themselves that a proposed solution is correct. To accomplish this students should be allowed to work on an assigned problem over several days and take their problem books home. At the end of the week students' answers can be corrected using the answers in *Problemoids: Math Mentor Level 5*.

By design the problems in *Problemoids* present challenges; consequently, they have the potential to be a source of frustration to students whose major goal is producing correct answers. Although some students are well aware when they are making progress and are perfectly satisfied that their work will eventually result in an answer, other students need reassurance that the problem can be solved and that the hints may provide assistance. Reassurance may be enough to motivate these students to continue working.

(5) ENCOURAGE STUDENTS TO READ AND EVALUATE SOLUTIONS PRESENTED ON SOLUTION CARDS.

A solution card should be posted immediately after students' answers have been checked. It is important that students have an opportunity to read the solution card and compare the suggested solution(s) with their own solution. This procedure provides for self-checking of the most complex part of students' work and enables students to learn problem solving strategies from the solutions whether or not they solved the problems correctly. Students should be encouraged to evaluate their solutions as correct, better than the authors' solution(s), or incorrect.

Students who believe they are close to a solution or at least on the right track may resist looking at a solution until they complete their own work. Such students might be allowed, within reason, more time to work on the problem before considering the authors' solution. A great deal of satisfaction and self-confidence can be gained from the knowledge that, "I did it myself!" Seeing a solution prematurely may cause resentment and frustration. However, since the solutions have instructional value, it would be inappropriate to allow a student to build up a backlog of unsolved problems.

In the event that any students abandon a problem in frustration, it is doubly important that they compare what they have done with the solution(s) provided by the authors. They may find that they actually were on the right track and that they had made real progress toward obtaining an answer, and, most importantly, they may understand what they could have done to solve the problem.

NOTE: It is important to distinguish the difference between an answer to a problem and a solution to a problem. Answers are just that—answers. They are often very short and usually provide no information about the strategies used to solve the problem. A solution is the process by which an answer to a problem is obtained. Solutions may be lengthy, and they provide information about the strategies used.

(6) NOTIFY STUDENTS OF THE LOOKING BACK FEATURE OF SOLUTIONS.

The solutions of many of the problems in *Problemoids* conclude with one or more LOOKING BACK suggestions. They have been included to help students find connections between problems, become more facile in solving problems, and experience the thrill of discovering something previously unknown.

There is a tendency, even among the best of problem solvers, to consider a solution finished once an answer has been obtained. Problem solvers lose a potentially valuable opportunity by stopping at the answer. LOOKING BACK can mean much more than checking and refining one's work. It can result in an increased awareness of the connections problems have with each other, greater facility with the use of problem solving strategies, and the discovery of new mathematical ideas. The solutions of many problems in *Problemoids* include LOOKING BACK suggestions to guide students in the pursuit of these goals.

(7) ENCOURAGE PARTICIPATION AND CONSIDER RECORDKEEPING.

The purpose of *Problemoids* is to reinforce, extend, and enrich the mathematics curriculum for students who have been identified as needing something more challenging than that offered in the typical curriculum. Students benefit from the program through active participation. Merely finding answers to the problems in *Problemoids* is not the objective of the program. Some educators might even argue that those who think about and work on problems for a while may benefit more than those who solve them quickly with little reflection.

Thus, teachers who implement *Problemoids* with students should encourage maximum participation from students. Evidence of participation may take the form of a great deal of scribbling in the student book, excitement about the program, increased interest in the program and/or in mathematics, student created or introduced problems, reluctance to stop working on a problem, conversations about problems, eagerness to see newly posted solutions, etc. In any case, the teacher, through expertise in teaching and personal acquaintance with the students, has a number of techniques to evaluate student participation.

Teachers may want to keep records on student participation. Since assigning a specific numerical grade to student participation is difficult, evaluations of "Satisfactory," "Needs Improvement," and "Unsatisfactory" may be sufficient. Some teachers may want to keep records on students' answers, but this procedure does not necessarily reflect the purpose of the program and it may have the negative effect of tacitly communicating to students that answers are the most important part of the program. Furthermore, the level of difficulty of problems generally increases with their order in the student book making it more difficult to evaluate progress only by checking answers.

BACKGROUND

ENRICHMENT

Problemoids has been carefully designed to meet the needs of intellectually gifted students and other students who demonstrate strong ability in mathematics. The *Problemoids* program complements normal mathematics instruction at the appropriate grade level by reinforcing, extending, and enriching each topic in the typical mathematics curriculum.

Problemoids reinforces the typical curriculum by requiring students to apply skills offered in the curriculum at their grade level. Participation in the program does not require that students know content in mathematics beyond their grade level although gifted children often do. Even if children have advanced mathematical skills, they will probably find the thinking challenges presented in *Problemoids* quite exciting.

Problemoids extends the typical curriculum by carefully introducing some new skills at more advanced levels. Such presentations have been designed to prevent confusion and the concomitant flood of questions. Teachers are not expected to assume responsibility for explication of new material.

The primary focus of *Problemoids* involves enrichment of the mathematics program. James Gallagher defines enrichment in *Teaching the Gifted Child* (pp. 76-77) as follows: Enrichment can be defined as the type of activity devoted to the further development of the particular intellectual skills and talents of the gifted child. These might be described as:

1. The ability to associate and interrelate concepts.
2. The ability to evaluate facts and arguments critically.
3. The ability to create new ideas and originate new lines of thought.
4. The ability to reason through complex problems.
5. The ability to understand other situation, other times, and other people, to be less bound by one's own peculiar environmental surrounding.

Problemoids satisfied all the criteria cited above. It involves students in challenging problem solving situations which facilitate learning problem solving strategies and stimulate searching for new problems as well as solving specific problems. The problems presented in this program are designed to be investigated through the use of problem solving strategies emphasized in the hints; their solutions demand a substantially higher level of thinking that the solutions of most standard textbook problems. Suggestions in *Problemoids* guide students to find new questions about problems and to create new problems with new questions. Through participation in these activities students acquire skills which they can apply to new situations.

EXPECTATIONS

It is expected that students who participate fully in the *Problemoids* program will show greater interest in mathematics and improved ability to solve new problems. Their interest may manifest itself in conversations about problems with peers or the teacher, the creation of new problems or new questions about problems, bringing new problems from other sources to school to share, willingness to work on assigned problems, reluctance to stop working on problems, excitement about the posting of solutions to assigned problems, etc. Teachers might recognize students' improved ability to solve new problems in improved organization of work, self-satisfaction with progress, a higher degree of flexibility in the use of strategies, increased self-confidence (risk taking) in problem solving, or expressed enjoyment of the program.

PROGRAM

The needs of intellectually gifted students are similar to those of other children in that they should be served on a regular and continuing basis. Students obtain maximum benefit by participating at all available levels of the *Problemoids* program; yet, students can participate at any level without prior exposure to the program. The different levels of the program coordinate presentation of problems and instruction in the use of strategies. *Problemoids* has the substance to stand as part of the permanent program for intellectually gifted students.

Problemoids: Math Mentor Level 5

Problemoids is a mathematics problem solving program especially designed to meet the needs of intellectually gifted students and students who might benefit from an enriched mathematics program more advanced than that provided in the standard curriculum. *Problemoids* engages students in enjoyable high level thinking about challenging problems and provides a stimulating educational opportunity for students to refine and increase their repertoires of problem solving strategies.

Solutions of problems require students to use the full spectrum of thinking skills in Bloom's Taxonomy with special attention to high level skills such as analysis, synthesis, and evaluation. The mathematical skills required of the students correspond to those stated in appropriate grade level curriculum guides of many states and school systems. Topics covered in the program extend, enrich, and reinforce topics typically covered in the mathematics curriculum (sets, number and numeration, operations, geometry, measurement, algebra, and probability and statistics).

The program includes student problem books and teacher materials. The student problem book, *Problemoids: Math Challenge Level 5*, contains fifty problems. Several problems enrich each topic in the typical curriculum (e.g., sets, geometry, etc.). The problems differ from those in the typical curriculum in that their solutions require one or more mental abstractions. Additionally, as many as four hints accompany each of the problems to assist students in solving the problems by directing them toward the use of problem solving strategies identified by George Polya in *How To Solve It*.

Teacher materials, *Problemoids: Math Mentor Level 5*, include problems, hints, answers, and solutions. solutions emphasize the efficacy of problem solving strategies through sequential development of each solution in accordance with its accompanying hints. In many instances the solution card contains more than one solution, which demonstrates to students that there is no single fixed, best, or correct process for solving a challenging problem. It also offers the opportunity for students to compare the thinking involved in their solutions with that in several different solutions, and it facilitates confirming correct solutions and detecting mistakes in incorrect solutions.

ANSWERS TO COVER STORY QUESTIONS

(1) The sum of the numbers in the tenth row is the tenth power of 2 or 1024.

(2) If the number of a row is a prime number, all the numbers except the 1's at the ends of the row are divisible by that prime. There are fourteen numbers in the 13th row including two 1's, so twelve numbers in the 13th row are divisible by 13.

(3) If the number of a row is one less than 2 raised to a power, all the numbers of that row are odd. Since 31 is one less than 32, the fifth power of 2, the 31st row has only odd numbers in it or 32 odd numbers.

APPENDIX I
THE STRATEGIES USED IN *PROBLEMOIDS*

The following is a list of the strategies used in the Hints and Solutions of *Problemoids*. An explication of the nature and use of these strategies follows
in Appendix II.

WHAT ARE YOU ASKED TO FIND?
WHAT INFORMATION IS GIVEN IN THE PROBLEM?
WHAT IS THE CONDITION?
HAVE YOU SEEN THIS TYPE OF PROBLEM OF A SIMILAR PROBLEM BEFORE?
MAKE A CHART or DRAW A DIAGRAM.
INTRODUCE AN ELEMENT.
SOLVE A SIMPLIFIED RELATED PROBLEM.
SOLVE PART OF THE PROBLEM.
USE ALL GIVEN AND IMPLIED INFORMATION.
SEARCH FOR A PATTERN.
USE THE TRIAL AND ERROR METHOD.
WORK BACKWARDS.
SOLVE AN ANALOGOUS PROBLEM.
CHANGE THE QUESTION.
CHANGE THE INFORMATION GIVEN IN THE PROBLEM.
CHANGE THE CONDITION.
RESTATE THE PROBLEM.
IF YOUR PLAN DOES NOT LEAD TO A SOLUTION, TRY A NEW PLAN.
LOOKING BACK.

Appendix II
DESCRIPTION OF THE STRATEGIES USED IN *PROBLEMOIDS*

WHAT ARE YOU ASKED TO FIND?
WHAT INFORMATION IS GIVEN IN THE PROBLEM?
WHAT IS THE CONDITION?

These three questions are directed mainly at understanding a problem. Three important parts of many problems are the unknown (what you are asked to find), the data (the information given in the problem), and the condition (which must be satisfied by the unknown and the data). It is important for a problem solver to distinguish all three parts of the problem and find the links between them. The unknown is the goal the problem solver works toward when solving a problem and should be kept in mind throughout the solution. The problem solver should explore various ways of reaching this goal. If the problem includes a large set of data, it is easy to forget or ignore some of the information during the solution. The problem solver should determine what information is given in the problem and examine whether all the stated information has been utilized. The condition of a problem may have several parts. The problem solver may produce a partial solution by separating the various conditions and considering each by itself or by dropping all but one condition and solving the problem with results.

HAVE YOU SEEN THIS TYPE OF PROBLEM OR A SIMILAR PROBLEM BEFORE?

This strategy serves as a starting point for investigating a problem and generating useful ideas for solving it. The object of this strategy is to find some feature or aspect of the problem which is familiar in some way and which might play a role in its solution. The answer to the similar problem may be the same as or closely related to the answer of the present problem, or the method used to solve the similar problem may be applied, perhaps with modifications, to solve the present problem.

MAKE A CHART or DRAW A DIAGRAM

This strategy is useful for organizing the conditions of or data related to a problem. A diagram may make the problem more understandable and may assist the problem solver in developing a viable strategy. A chart can improve the problem solver's efficiency in going through the steps of solving a problem and aid in the process of planning the next step. Charts also provide an excellent means of observing patterns.

INTRODUCE AN ELEMENT

There are several ways to introduce an element into a problem. In a geometry or measurement problem it may be productive to introduce an element by assigning a measure to some part of a diagram (for example, a length or an area) or to introduce a line segment or another geometric element in a diagram. In other problems there might be another type of element not mentioned in the statement of the problem which could be introduced in an effort to make the solution of the problem simpler.

SOLVE A SIMPLIFIED RELATED PROBLEM

There are several ways a problem can be simplified: large numbers can be replaced by smaller numbers, a variable can be replaced by a specific numerical value, a complicated diagram can be replaced by a simpler one, etc. The key is to find a simplification which makes the problem easier to solve but makes no drastic change in the basic structure of the problem. The intention is that the solution to the simpler problem will be applied or adapted to apply to the solution of the original problem.

SOLVE PART OF THE PROBLEM

Frequently it is possible to separate a problem into several component parts. In such cases substantial progress toward the solution can be made by a problem solver who considers the problem one part at a time. In fact, the solution to one part of a problem might produce a method which can be used to solve the other parts of the problem.

USE ALL GIVEN AND IMPLIED INFORMATION

At times, particularly when a problem has a large set of data or conditions, some information in the statement of the problem may be ignored or forgotten as one solves the problem. In these instances all that may be needed to complete the solution is some information already stated in the problem. More frequently what one needs to make progress in a solution is not actually stated in the problem but is something that can be deduced from stated information or from information previously deduced.

SEARCH FOR A PATTERN

Searching for a pattern may be used in conjunction with one or more other strategies such as make a chart, use the trial and error method, or solve part of the problem. Searching for a pattern involves generating data (if necessary), organizing the data, and then analyzing the data for regularities or patterns. Any observed pattern which seems to be potentially useful needs to be tested. The test might be to generate more data and check whether they satisfy the observed pattern or to produce a logical reason why the pattern works.

USE THE TRIAL AND ERROR METHOD

The use of the trial and error method in the early stages of a solution often leads to a valuable understanding of the problem. This strategy also can be applied at other stages of the solution, but, whenever it is applied, it is important to realize that most trials do not result in an answer to the problem being solved. By making trials the problem solver hopes to form a successively clearer idea about the answer. When a trial fails to yield the correct answer, the problem solver should ask himself, "What information can I salvage from the failed trial? What would be a reasonable next trial?"

WORK BACKWARDS

When employing the work backwards method, the problem solver focuses not on the data or conditions of the problem but on what must be proved or what must be found. The problem solver might ask, "What do I need to show that this is true?", or "What do I need to know in order to find this?" or "If I had this, what else would be true?" (The "this" in these questions refers to what must be proved or found.) The object of the work backwards method is to answer such a question by citing something already stated or easily deduced from what is stated in the problem. Often it is necessary to work backwards for several steps before completing a solution.

SOLVE AN ANALOGOUS PROBLEM

If two problems are similar, they are analogous. The level of similarity may be rather vague, quite precise, or somewhere in between. The object of this strategy is to find a less difficult problem which is sufficiently similar to the problem being solved that its answer or solution can be adapted to the problem being solved. The analogous problem may be a familiar problem with a known solution or it may be an unfamiliar problem which is similar to the present problem in what must be found or proved, stated information, or condition(s).

CHANGE THE QUESTION

At times the same data and conditions stated in a problem can lead to several different questions or questions worded in different ways. When using a particular question does not yield a solution easily, it may be advantageous to consider using an alternate question. The answer to the alternate question may be directly related to the original question, or the solution to the alternate question may be adopted or modified for use in the original problem.

CHANGE THE INFORMATION GIVEN IN THE PROBLEM

Problem solvers might expedite their solution of a problem by changing some of the information given in the problem. This involves changing some of the data of the problem.

CHANGE THE CONDITION

A condition of a problem is a statement or a relationship that the data must satisfy. Manipulation of the condition(s) of a problem may produce a problem which is more easily solved that the original problem or may lead directly to a solution of the original problem.

RESTATE THE PROBLEM

Occasionally problem solvers may restate a problem in their own words to make the problem more understandable and expedite its solution. However, the strategy of restating the problem is broader than this. Any variation of the statement, question, information given, or condition of a problem is a restatement of the problem. The object of restating the problem is to produce a variation which is equivalent or very similar to the original problem but which is simpler to solve.

IF YOUR PLAN DOES NOT LEAD TO A SOLUTION, TRY A NEW PLAN

The fact that using a particular strategy to solve problems is often fruitful does not guarantee that it always will be productive or expedient to use that strategy. Problem solvers should be flexible rather than bound to a single plan or strategy. Problem solvers should be flexible rather than bound to a single plan or strategy. They must judge when a chosen plan is not productive, abandon it, and try a new plan.

LOOKING BACK

After completing the solution of a problem there are several techniques the problem solver can use to look back at the solution. Some of these techniques are employed to check the validity of the answer or the efficiency of the solution. For example, the problem solver can check calculations for accuracy, check to make sure that each step of the solution is correct, check to see that the answer fits all the conditions of the problem, or consider whether there may be a better solution to the problem.

Looking back also can be performed in a manner that resembles how professional mathematicians explore problems and solutions to make new mathematical discoveries. Mathematicians often examine the relationship between a problem just solved and previously solved problems. Problem solvers can examine their solution for some new mathematical ideal and investigate the implications of the ideas, or they can vary the original problem and explore what effect the change has on the solution. By using these techniques, problem solvers create the opportunity to make an exciting discovery.

Appendix III
CLASSIFICATION OF PROBLEMS ACCORDING TO MATHEMATICAL TOPIC

● = primary classification
★ = secondary classification

		Sets	Number Numeration	Operations	Algebra	Geometry	Measurement	Probability & Statistics
1.	Yin and Yang	●						
2.	Inflation		●	★				
3.	Dribbles		●			★		
4.	A New Game Plan			★	●			
5.	What's The Difference?		★	●				
6.	Where There's a Will, There's a Way					●	★	
7.	Bit Buy Byte			★	●			
8.	Confronting the Clock		●				★	
9.	Candydextrous			★	★		●	
10.	Aw Sum Gauss		●	★				
11.	The Muse Family	●						
12.	A Ball with the Bearings			●	★		★	
13.	Sexagenarian					●		
14.	The Toothpick Fairy					●		
15.	Lots of Dots		●					★
16.	A Liter Expedition			★			●	
17.	Splitting the Spoils		★					●
18.	Sweet Enigma			★	●			
19.	Crack the Cube					●		
20.	Fifty Point Question		●			★		★
21.	A Little More			●	★			
22.	Check'er Balance				●			
23.	Caveat Emptor			★		★	●	
24.	M & M's		●					
25.	Hoppin' To Win			●				
26.	The Case of the Quick Bookcases		★	★	●			
27.	Hex Symbol					●		
28.	Pause for Claws			★	●			
29.	The Wandering Bishop		★					●
30.	Computing the Odds		●					
31.	Distributing the Wealth			●	★			
32.	Friendly Finish				●		★	
33.	Getting Even	●		★				
34.	An Almost Perfect Place		●					★
35.	Chew on This One					★	●	
36.	A Walking Conundrum				●		★	
37.	Hundreds Censored	★		●				
38.	Try Your Luck					●		
39.	Quick Candlestick				★		●	
40.	Partly Popular			★	●			
41.	An Odd Gift		●					
42.	Fair Ball							●
43.	Creating Cranapple			★			●	
44.	Relatively Unknown		★	●				
45.	Name Dropping		●					★
46.	Parsimonious Purchaser	●						
47.	Installation Free					●		
48.	Quintillion Plus		●	★				
49.	Three-Quarter Dime Jingle					●		★
50.	No Heads Is Better Than One							●

1. YIN AND YANG

A traveler planning a journey through unfamiliar country know that he will reach a fork in the road at which he will not know which way to go. Two people who live at the fork customarily give directions to strangers, but one always lies and the other always tells the truth. Since they don't like talking, they allow travelers to ask only one of them one question about directions. The traveler doesn't know which inhabitant tells the truth and which lies, so what question should he ask and which inhabitant should he ask?

Hint 1
Use the trial and error method to examine different cases and different types of questions.

Hint 2
Work backwards assuming that you are questioning the liar and construct a question such that his response will yield useful information.

ANSWER: Since the traveler cannot tell who is the liar and who tells the truth, there are two cases: one in which he asks the inhabitant who tells the truth and one in which he asks the liar. Thus the problem becomes the construction of a question which will yield useful information in either case. The case of asking the truthful inhabitant is simpler to satisfy. The object now is to construct a question which will force the liar to lie twice in such a manner that one lie will negate the other, or to tell the truth. Such a question might be, "If I were to ask you if I should take the left fork, would you say yes?"

2. INFLATION

Make four 9's equal 100. Evaluate your progress using the criteria below.
So-so: two solutions
Good: three or four solutions
Very Good: five or six solutions
Excellent: seven or eight solutions
Incredible: nine or ten solutions

Hint 1
Use the trial and error method and mathematical operations that you know to make numbers with four 9's.

Hint 2
Solve a simplified related problem. For example, make three 9's equal ten.

Hint 3
Solve another simplified related problem. Make two 9's equal one.

ANSWER: Some of the possible solutions are:

1) $99 + 9/9$
2) $99^9/_9$
3) $99/.99$
4) $9 \times 9/.9 \times .9$
5) $99 + \sqrt{9}/\sqrt{9}$
6) $9 + (9 \times 9)/.9$
7) $9 \times 9/.9/.9$
8) $9/.9 \times 9/.9$
9) $99 + .9/.9$
10) $99 + \sqrt{.9}/\sqrt{.9}$

Hint #2 $9/9 + 9 = 10$
Hint #3 $9/9 = 1$

3. DRIBBLERS

The Little Bounce Sporting Goods Company orders its winter supply of basketballs in the summer and stores them until they are needed. This year the delivery arrived when no one was available to put it in the warehouse. The cube-shaped boxes of basketballs were left outside in a neat stack seven high, seven wide, and seven deep. It rained unexpectedly that night and the store manager discovered that all the boxes that were exposed to air were ruined. He angrily called his supplier and demanded replacements for the basketballs in ruined boxes.

How many replacements do you think he ordered?

Hint 1
Solve part of the problem. For example, how many boxes were in the stack?

Hint 2
Solve part of the problem. For example, how many boxes were not ruined.

Hint 3
Draw a diagram of the stack or make a model of it.

Hint 4
Solve a simplified related problem, for example, how many boxes would be ruined under similar conditions in a stack three high, three deep, and three wide?

ANSWER: 193

4. A NEW GAME PLAN

Peter's dad looked at the money Peter was counting. "You certainly have saved a lot of money for that game you've been wanting to buy; I guess that you have enough to buy it now," he said.

"Yes, I do," Peter exclaimed. "I was really lucky today. I found five dollars on my way home from school."

"That was lucky," his father said. "Now you have five times as much as you would have had if you had been unlucky and lost five dollars."

How much money did Peter have before fortune struck?

Hint 1
What are the conditions?

Hint 2
Use all given and implied information. How can you limit the possible answers?

Hint 3
Use the trial and error method trying to make each attempt closer to the correct answer.

ANSWER: $7.50

5. WHAT'S THE DIFFERENCE?

The subtraction problems below may prove to be more interesting than they look. The letters indicate correct solutions in basic arithmetic, and they are also arranged so that a digit can be substituted for each letter. What substitutions do you suggest?

$$\begin{array}{r} \text{TWO} \\ -\text{ONE} \\ \hline \text{ONE} \end{array} \qquad \begin{array}{r} \text{FIVE} \\ -\text{FOUR} \\ \hline \text{ONE} \end{array}$$

Hint 1
Restate the problem. Subtraction problems may be somewhat easier to solve if restated as addition problems, for example.

$$\begin{array}{r} \text{ONE} \\ +\text{ONE} \\ \hline \text{TWO} \end{array} \qquad \begin{array}{r} \text{ONE} \\ +\text{FOUR} \\ \hline \text{FIVE} \end{array}$$

Hint 2
Use all given and implied information. The problem states that letters have certain numerical values. What, for example, are limits that you might place on the value of some letters?

Hint 3
Now that you have established some limits, use the trial and error method to test values for letters.

ANSWER: Some possible solutions are:

$$\begin{array}{r} 572 \\ -286 \\ \hline 286 \end{array} \quad \begin{array}{r} 3496 \\ -3210 \\ \hline 286 \end{array} \quad \begin{array}{r} 472 \\ -236 \\ \hline 236 \end{array} \quad \begin{array}{r} 9516 \\ -9280 \\ \hline 236 \end{array}$$

6. WHERE THERE'S A WILL, THERE'S A WAY

The Peterson quadruplets inherited an L-shaped plot of land from the estate of a distant relative. Since the quadruplets were identical in size and shape, they desired to split the land into four pieces of the same size and shape. How can they do so if the long sides of the plot are twice as long as the short side and if all pairs of consecutive sides are perpendicular?

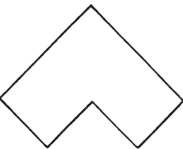

Hint 1
Solve a simplified related problem. For example, divide an equilateral triangle into four pieces of the same size and shape.

Hint 2
Work Backwards using the area of each of the four proposed pieces.

ANSWER:

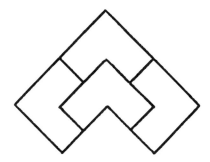

7. BIT BUY BYTE

One afternoon Jason went to visit a new friend, Tom, at his home. When they entered Tom's room, Jason saw his friend's microcomputer and exclaimed, "Wow! You have your own computer; I'll bet that cost a fortune!"

Tom replied, "Actually, I earned the computer by working part-time during the fall season at the store that sells them. The owner agreed to give me a computer and $240.00 for working 18 weeks. I got so busy with schoolwork that I had to quit after 12 weeks, so he gave me $60.00 and the computer."

What was the value in dollars of Tom's computer?

Hint 1
Use the trial and error method to make successively closer guesses of the value of the computer.

Hint 2
Change the question to "How much would Tom have been paid for the six weeks he did not work?"

Hint 3
Use all given and implied information to write an equation.

ANSWER: $300.00

8. CONFRONTING THE CLOCK

Hector runs around a circular track once every 90 seconds. Earl, starting at the same spot but running in the opposite direction, meets Hector every 40 seconds. How long does it take Earl to run around the track?

Hint 1
Draw a diagram.

Hint 2
Use all given and implied information. For example, does Earl take more time or less time than Hector to run one lap?

Hint 3
Change the question. How far does Hector run in 40 seconds?

Hint 4
Use all given and implied information. If you know how far Hector runs in 40 seconds, what portion of the lap does Earl run in that time.

ANSWER: 72 seconds

9. CANDYDEXTROUS

Little Nina Martin loved candy kisses so much that she taught herself how to unwrap the delicious chocolate morsels one-handed with each hand. She could unwrap two at a time—one in each hand! However, since she was naturally left-handed, she was not as efficient using her right hand. In fact, her right hand was only two-thirds as fast as her left hand.

One cold February day Nina's father gave her a big bag of candy kisses. Nina began unwrapping the candies with her left hand. Thirty minutes later, her right hand joined in the task. Thirty minutes after that, every one of the candy kisses had been unwrapped. Nina had not eaten one yet. She put them into a bowl so that she could share them with her brothers and sisters.

When her sisters and brothers arrived, Nina told them how she had gone about unwrapping the candies and posed this problem:

"If I had used both hands at the start," she said, "and if my right hand had stopped when half the candies were unwrapped, how much time would it have taken my left hand to unwrap the other half of the candies?"

Nina's brother, Billy, quickly answered, "30 minutes!"

"No, Billy, that's not correct," Nina replied.

What is the answer to Nina's question?

Hint 1
Change the question. How much time would it take Nina's left hand working alone to unwrap the bag of candies?

Hint 2
Use all given and implied information. How much time would it take Nina's left hand to unwrap the candies which Nina's right hand unwrapped in thirty minutes?

Hint 3
Use all given and implied information. How long did Nina's left hand work as she unwrapped the candies and put them into the bowl?

ANSWER: 40 minutes

10. AW SUM GAUSS

History books record early and amazing events in the life of Carl Friedrich Gauss, the famous German mathematician sometimes called "The Prince of Mathematicians." One such event is about the time his teacher gave him a long sequence of numbers to add in which the difference between consecutive numbers was the same each time. Gauss solved the problem almost as soon as his teacher finished stating it.

Perhaps you can find as easy method of solving a similar problem. What is the sum of: $1 + 2 + 3 \ldots + 100$?

Hint 1
Solve a simplified related problem. For example, list the numbers from 1 to 10 and try to find a simple way of adding them.

Hint 2
Make a chart by adding the numbers in several sequences of small integers and search for a pattern in your chart.

ANSWER: $100/2 \times 101 = 5050$

11. THE MUSE FAMILY

Some members of the Muse family are preparing for a trip. A neighbor knows that they are related to each other in the following manner: three daughters, two sons, two mothers, two fathers, two granddaughters, one grandson, one grandfather, one great-granddaughter, one grandmother, and one great-grand-father. Interestingly enough, they all fit into their standard-size car comfortably. How is this possible?

Hint 1
Draw a chart. Construct a family tree showing possible relationships between family members.

Hint 2
Solve a simplified related problem, for example, "Fishy Canoe." Two fathers and two sons go fishing in a canoe. Each person catches one fish and they have three fish altogether. How is this possible?

Hint 3
Solve part of the problem. Determine the number of males in the car.

ANSWER: Six people go in the car: a man, his son, his son's son, his daughter, his daughter's daughter and granddaughter.

12. A BALL WITH THE BEARINGS

Suppose some ball bearings each weigh 3/5 pound and 1/16 of a ball bearing. In pounds, how much would 25 of the ball bearings weigh?

Hint 1
Draw a diagram of a balance scale and the items in the problem.

Hint 2
Solve a simplified related problem. For example, if a brick weighs half a pound and half a brick, how much does one brick weigh?

Hint 3
Solve part of the problem. How much does one ball bearing weigh?

ANSWER: 16 pounds

13. SEXAGENARIAN

"Age is something I'm proud of," said grandfather. "I would be as old now as six times my age six years from now less six times my age six years ago if I were six years older than I am."

How old is grandfather now?

Hint 1
What are the conditions? State the conditions in a manner that is clear to you.

Hint 2
What information is given in the problem? Use this information to write an equation.

Hint 3
Use the trial and error method and the conditions of the problem to make successively closer guesses of grandfather's age.

ANSWER: 66

14. THE TOOTHPICK FAIRY

(A) Make six four-sided figures with the same size and shape by moving six of the toothpicks in the figure below.
(B) Remove five toothpicks from the figure below to form five triangles.
(C) Remove six toothpicks from the figure below to form five triangles.

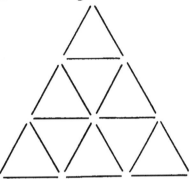

Hint 1
Solve part of the problem. What sorts of four-sided figures can be produced by moving or removing toothpicks?

Hint 2
Work backwards.

Hint 3
Use the trial and error method. What types of four-sided figures can you make?

ANSWER: Some of the many possible answers are illustrated here.

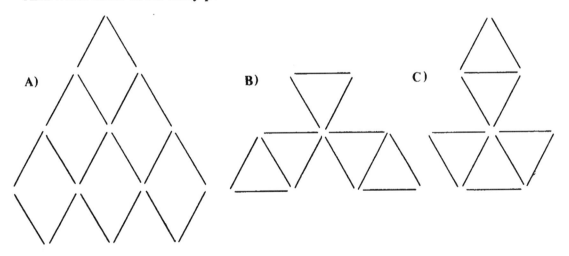

15. LOTS OF DOTS

The numbers 1, 3, and 6 are sometimes called the first three triangle numbers because they are associated with the number of dots in the following arrays of dots:

```
                        .
              .        . .
    .        . .      . . .
    1         2          3
```

What is the fiftieth triangle number?

Hint 1
Have you seen this type of problem or a similar problem before, for example, Aw Sum Gauss?

Hint 2
Solve a simplified related problem. Find the fourth triangle number.

Hint 3
Search for a pattern. Find the 5th, 6th, and 7th triangle numbers and see if you can use the results to find the 50th triangle number.

Hint 4
Make a chart to help you find a pattern.

ANSWER: The fiftieth triangle number is 1,275.

16: A LITER EXPEDITION

On their recent camping expedition, Lewis and Clark encountered a challenging problem. They took three jugs with them on their expedition; the jugs had capacities of 10 liters, 7 liters, and 3 liters, respectively. Clark went to fetch some water in the 10 liter jug. When he returned, Lewis said, "For cooking our meal tonight we need two equal portions of water—5liters each. How are you at estimating?

Clark boasted, "We do not need to estimate. I am certain that we can divide the water into two equal portions just by using our three jugs. In fact, I bet I can do it in ten pourings and not spill a drop."

Should Clark have explored his thinking further? Can he accomplish the task in just ten pourings? What is the least number of pourings required for the task?

Hint 1
Change the information given. Suppose the jug capacities were 10, 6, and 1 liter(s).

Hint 2
Work backwards assuming that you have two 5 liter portions.

Hint 3
Use trial and error keeping a record of your attempts.

ANSWER: Clark can accomplish the task in ten pourings:

START	1st	2nd	3rd	4th	5th	6th	7th	8th	9th	10th
10,0,0	7,0,3	7,3,0	4,3,3	4,6,0	1,6,3	1,7,2	8,0,2	8,2,0	2,5,3	5,5,0

However, he could do it in nine pourings:

START	1st	2nd	3rd	4th	5th	6th	7th	8th	9th
10,0,0	3,7,0	3,4,3	6,4,0	6,1,3	9,1,0	9,0,1	2,7,1	2,5,3	5,5,0

17. SPLITTING THE SPOILS

Pam finds a bag with 30 marbles. She decides to share them with her friends Juan and Leon. In how many different ways can she distribute the marbles so that each of the three children gets at least one?

Hint 1
Use all given and implied information. For example, what is the greatest number of marbles any one boy can have?

Hint 2
Make a chart and search for a pattern.

Hint 3
Change the information given. For example, suppose Pam shares the marbles only with Juan.

Hint 4
Solve a simplified related problem. For example, reduce the number of marbles.

ANSWER: 406 ways

18. SWEET ENIGMA

Three enterprising children decided to sell boxes of candy door-to-door. Each started the month of July with the same number of boxes of candy and each sold 60 boxes by the end of July. At the end of July the three children together had as many boxes of candy as each had originally. How many boxes of candy did each child have at the beginning of July?

Hint 1
What are the conditions of the problem?

Hint 2
What are you ask to find?

Hint 3
Use all given and implied information. Write an equation.

ANSWER: 90 boxes

19. CRACK THE CUBE

Below are drawings of the same cube in three different positions. How many spots are on the bottom (consider the side with six spots as the top) of the cube in position number one?

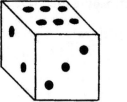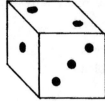

Hint 1
Draw a diagram of the cube as if it were unfolded.

Hint 2
Use all given and implied information about the relationships among the adjacent faces.

Hint 3
Is each step of your plan correct? Compare your diagram with the given information.

ANSWER: Two dots.

20. FIFTY POINT QUESTION

How many diagonals are there in a convex polygon with 50 sides?

Hint 1
Solve simplified related problems and made a chart of the numbers of diagonals in polygons with fewer sides.

Hint 2
Search for a pattern in your chart.

Hint 3
Have you seen this type of problem or a similar problem before, for example, aw sum gauss or splitting the spoils? How did you solve the similar problem?

ANSWER: 1175 diagonals.

21. A LITTLE MORE

Find the numerical value in lowest terms of the expression below.

$1/1{\times}2 + 1/2{\times}3 + 1/3{\times}4 + \ldots + 1/99{\times}100$

Hint 1
Make a chart by solving some simplified related problems and by organizing the results.

Hint 2
Search for a pattern in your chart.

ANSWER: 99/100

22. CHECK 'ER BALANCE

The scales below balance in the first two diagrams. In the third diagram how many checkers does it take to balance the pocket calculator?

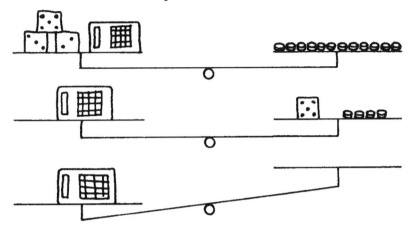

Hint 1
Introduce an element in the second diagram. What could you add to both sides of the balance scale, keeping it balanced, to make one side of it like the first diagram?

Hint 2
Change the question to, "How many checkers are required to balance one die?"

ANSWER: 6 checkers

23. CAVEAT EMPTOR

Meg and Charlie usually order one mushroom and sausage pizza at the C & E Pizzeria. This pizza has a diameter of 50 centimeters and costs $5.00. Last week, however, C & E had a special: three 25 centimeter pizzas for $5.00.

"Let's try the special," Charlie said, expecting to gorge himself with the wonderfully oily morsels of the C & E specialty. "This place will go broke if they keep this special for very long."

After the meal, however, Meg complained, "I think this pizzeria is making more money than ever!"

Why was Meg so upset?

Hint 1
Draw a diagram comparing the sizes of the pizzas.

Hint 2
Solve an analogous problem. Suppose the pizzas were square shaped; what would be the effect of halving the length of one side?

Hint 3
Do a bit of research. Find a formula for calculating the area of a circle.

ANSWER: Meg was upset because she had to pay the same amount for the special but received only three-fourths as much pizza as she would ordinarily get.

24. M & M's

Sam's uncle owned a plant that made M & M's. While visiting the plant, Sam spotted 150 sample bags of M & M's, and he asked his uncle if he could take them to school to share with his classmates.

His uncle replied, "Well Sam, each of the sample bags has a different number from one to 150, and each bag contains the same number of M & M's as its number. I don't want you and your classmates to eat too many sweets, so you may take every third, fifth, and seventh bag of the 150 bags."

Sam thanked his uncle and began to collect the bags. How many M &M's did Sam have to share with his classmates?

Hint 1
Solve a simplified related problem. For example, suppose there were only 15 sample bags. How many M & M's would Sam have?

Hint 2
Solve part of the problem. For example, how many M&M's will taking every third bag provide?

Hint 3
Have you seen this type of problem or a similar problem before?

ANSWER: Sam had 6109 M & M's to share with his classmates.

25. HOPPIN' TO WIN

This year the members of the 4-H Club plan to entertain the preschool children in their community with an animal race. They decide to mark a straight race course one hundred feet in length. The winner will be the first animal to run to the end and back.

The racing dog makes three-foot jumps, the goat leaps eight feet per jump, and the rabbit jumps two feet at a time. The rabbit, however, can jump three times each time the dog jumps twice and four times each time the goat jumps once.

Of course the members of the 4-H Club know which animal will win, but they plan to ask their young friends whether they can pick the winner before the race.

Which animal do you think will win? Which will finish second? Explain your answer.

Hint 1
Draw a diagram of the race course and the jumps of the animals.

Hint 2
Solve part of the problem. How many jumps will it take the dog to reach the one hundred foot mark?

Hint 3
Use all given and implied information. How far must the dog travel to complete the entire race?

ANSWER: The rabbit will win the race, and the dog will finish second.

26. THE CASE OF THE QUICK BOOKCASES

The Omega Office Furniture Company is making bookcases for the new library at the state university. According to the contract, Omega should manufacture 24 bookcases per day in order to deliver the bookcases on time. However, new equipment and procedures have improved productivity at Omega and the company produces one more bookcase per day than was specified in the contract. The bookcases are finished and delivered two days early. How many bookcases did the library receive?

Hint 1
Change the question. How many days did Omega work producing 24 bookcases a day?

Hint 2
Use the Trial and Error method, guessing how many days Omega was to work producing 24 bookcases a day.

Hint 3
If your plan does not lead to a solution, try a new plan. Use all given and implied information to construct an equation.

ANSWER: 1,200 bookcases.

27. HEX SYMBOL

While experimenting with cutouts of geometric figures, Rick was surprised when he found two irregular hexagons of the same shape but different sizes which he could not, fit together no matter how hard he tried. That is, he could not make the larger one cover the smaller one completely. What were the shapes of these vexing hexagons?

Hint 1
What information is given in the problem? What can irregular hexagons look like?

Hint 2
Use the trial and error method to construct possible solutions.

ANSWER: There are no answers among the irregular convex hexagons; however, there are many answers among the irregular hexagons which are not convex. For example,

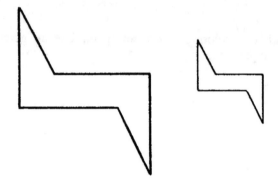

28. PAUSE FOR CLAWS

Catie asked her mother if she could raise some pets. "Melissa has some gerbils and some pretty little canaries," she said.

"That's quite a menagerie!" said Mrs. Randolph. "How many pets does Melissa have?"

"They have twelve heads and forty feet," replied Catie, who was very fond of puzzles.

How many gerbils and how many canaries does Melissa have?

Hint 1
Use all given and implied information. For example, the problem states, "They have twelve heads..." How many pets does this information imply that Melissa has?

Hint 2
Use the trial and error method to make successively closer guesses as to how many of each type animal Melissa has.

ANSWER: 8 gerbils and 4 canaries

29. THE WANDERING BISHOP

A rather serious young chess player sits contemplating a problem that he thought of during a game with his brother. He imagines that there is only one bishop on the board and he wants to move it from some one of the white squares on the first row to a white square on the eighth row. He knows that there will be many different routes by which he can move the bishop from a starting square to a finishing square (moving diagonally with no backwards moves), and he wants to find the largest possible number of routes and identify the starting and finishing squares.

What solution will satisfy this young man?

One sample route:

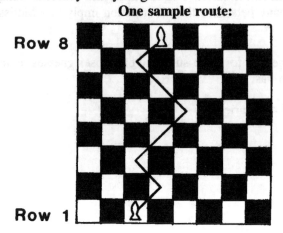

Hint 1
What are you asked to find? Does it make any difference which square you start with and finish on?

Hint 2
Solve a simplified related problem. Use the trial and error method to determine the largest possible number of routes from the first to the fourth row and which are the beginning and ending squares.

Hint 3
Search for a pattern in your work.

ANSWER: 35 routes. The starting and finishing squares are indicated below:

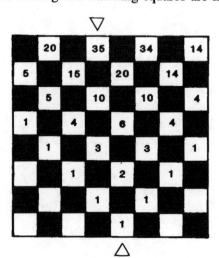

30. COMPUTING THE ODDS

Compute the sum of the first one hundred positive odd integers.

Hint 1
Have you seen this type of problem or a similar problem before, for example, Aw Sum Gauss?

Hint 2
Make a chart of sums of consecutive odd integers.

Hint 3
Search for a pattern in your chart.

ANSWER: 10,000.

31. DISTRIBUTING THE WEALTH

Three friends, Ramon, Karen, and Laverne, try an experiment with the coins they are carrying. Ramon gives Karen as many cents as she already has and Laverne as many cents as she already has. Then Karen gives Ramon as much as he now has and Laverne as much as she now has. Finally, Laverne gives Ramon and Karen as much money as each then has. They are surprised to find that each now has sixty-four cents. How much did each person start with?

Hint 1
Use all given and implied information. If each had sixty-four cents after the third set of transactions, how much did each have after the second set of transactions?

Hint 2
Make a chart representing the amounts each person had before and after each transaction.

Hint 3
Solve an analogous problem. Suppose a farmer's daughter started to market with some eggs. She met three customers at different places along the way and each brought half her eggs plus half an egg. She sold all her eggs to these three customers. How many eggs did she start with?

ANSWER: Ramon started with $1.04, Karen with $.56, and Laverne with $.32.

32. FRIENDLY FINISH

Rachel and Laurie have a bicycle race from their hometown, Bethel, to the town of Colebrook which is 40 miles away. Rachel is the faster cyclist in this race. She reaches Colebrook and immediately starts the return trip continuing at the same speed until she meets Laurie four miles from Colebrook. They stop and determine that Laurie's average speed was two miles per hour slower than Rachel's. How fast did each girl travel?

Hint 1
Change the question. How far did each girl cycle?

Hint 2
Use all given and implied information. Compare Rachel's average speed with Laurie's.

Hint 3
Use all given and implied information. Determine the length of time the girls travelled.

ANSWER: Rachel's average speed was 11 miles per hour and Laurie's was 9 miles per hour.

33. GETTING EVEN

Dave, Johnny, and Tom were very good friends. After Tom's wife died, his niece kept house for him. Dave, also a widower, lived with his daughter. When Johnny got married, he and his wife suggested that they all live together. Each person in the group was to contribute $100.00 on the first of the month for household expenses, and any extra money at the end of the month was to be equally divided.

Household expenses totaled $368.00 for the first month. When the extra money was distributed, each person received a whole number of dollars. How much did each person receive and why?

Hint 1
Have you seen this type of problem or a similar problem before, for example, the Muse Family?

Hint 2
Use all given and implied information. For example, what do you think the greatest and least possible numbers of people involved are?

Hint 3
If your plan does not lead to a solution, try a new plan. Did you guess that there were six people? If you did, look carefully to see if a different number of people could be involved.

ANSWER: Johnny's wife was Dave's daughter and Tom's niece, thus there were only four people. $400 was contributed, $368 spent, and each person received $8 in the distribution.

34. AN ALMOST PERFECT PLACE

At the turn of the century, the population of the village of Middlesboro reached 100, a perfect square. The 1960 census found that the population had grown and was again a perfect square. The 1970 census showed a population increase of 100 since 1960, but the population was now four more than a perfect square. By the 1980 census, after another gain of 100, the population was again a perfect square. What was the population of Middlesboro in 1960, 1970, and 1980?

Hint 1
What are you asked to find?

Hint 2
Change the condition. Suppose the population were a perfect square at each census. By how much should the population have grown each decade?

Hint 3
Use all given and implied information. Find three perfect squares such that the first two differ by 96 and the second two differ by 104.

Hint 4
Make a chart of consecutive perfect squares and search for patterns.

ANSWER: The population was 529 in 1960, 629 in 1970, and 729 in 1980.

35. CHEW ON THIS ONE

The toothpick fairy loves to construct geometric figures using the toothpicks she finds in the strangest places. In the construction below, the toothpick fairy has used thirty toothpicks and has made two rectangles, one of which has an area which is half that of the other figure.

The toothpick fairy enjoys watching people trying to use the same thirty toothpicks to form two four-sided figures, one having an area which is one-third that of the other. Construct two such four-sided figures.

Warning! The Toothpick Fairy gets very angry whenever anyone breaks a toothpick. You must solve this problem without damaging the toothpicks.

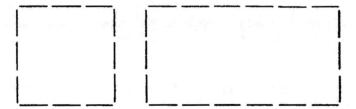

Hint 1
Draw a diagram.

Hint 2
Use the trial and error method to investigate the possible dimensions of the two figures.

Hint 3
Work backwards, construct one figure and then determine what the dimensions of the other figure might be.

Hint 4
What are you asked to find? Do the figures have to be rectangles?

ANSWER: One possibility:

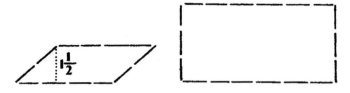

36. A WALKING CONUNDRUM

A commuter is in the habit of arriving at her suburban Connecticut train station each evening at 6:00 P.M. Her husband always meets the train and drives her home. One day she takes an earlier train arriving at the station at 5:00 P.M. The weather is quite pleasant, so, instead of telephoning home, she starts walking along the route always taken by her husband. They meet somewhere along the way, she gets into the car, and they drive home arriving ten minutes earlier than usual. Assuming that her husband always drives at a constant speed and that on this occasion he left just in time to meet the 6:00 P.M. train, how long had the wife been walking before being picked up?

Hint 1
Draw a diagram.

Hint 2
Use all given and implied information. For example, what is the meaning of the arrival time which is 10 minutes earlier than usual?

Hint 3
Use the trial and error method to experiment with different departure and return times for the car trip and search for a pattern.

Hint 4
Change the question. For example, at what time did the husband meet his wife?

ANSWER: 55 minutes.

37. HUNDREDS CENSORED!

Consider the integers 1, 2, 3, ... , 99. Find a set of integers in this sequence such that the sum of any of them does not equal 100. What is the largest such set that you can find?

CHALLENGE: Find another solution.

Hint 1
Change the condition from, "... the sum of them does not equal 100," to, "..the sum of two or more is greater than 100."

Hint 2
Make a chart to keep a record of your work.

ANSWER: There are many sets such that the sum of any of the integers does not equal 100. A common aspect of the largest such sets is that they must not contain more than 50 integers from the sequence 1, 2, 3, ..., 99, or else some combination will add to 100. One method of approaching the solution involves choosing integers so that the sum of any two is greater than 100; for example, the sequence 50, 51, 52, ..., 99. A few other solutions are:
the sequence 50, 51, 52, ..., 99 except replace 51 with 49,
the sequence 50, 51, 52, ..., 99 except replace 52 with 48,
the sequence 50, 51, 52, ..., 99 except replace 53 with 47, etc.

38. TRY YOUR LUCK

Long ago there was a king with seven children. He wanted to give each of them a gift and he also wanted to determine which of them would be the smartest successor to his throne. He and the blacksmith conspired to create a contest which would satisfy both of his desires. The blacksmith made a horseshoe of solid gold with seven nail holes in it. The king, concealing his secondary motive, told his children that if one of them could decide how to cut the horseshoe into seven pieces with a nail hole in each one using only two strokes of a sword, he would give each of them a piece for a necklace. How would you solve this puzzle? Can you also determine the maximum number of pieces that could be obtained using two cuts and disregarding nail holes?

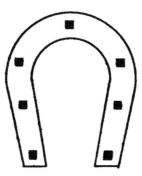

Hint 1
Draw a diagram.

Hint 2
Solve a simplified related problem. Using two cuts of the sword, cut the necklace pictured below into six pieces, each containing one jewel.

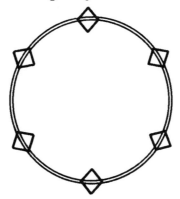

Hint 3
Introduce and element. Rearrange the pieces of the horseshoe (or necklace) before the second cut.

ANSWER: Notice: The pieces must be rearranged before the second cut is made.

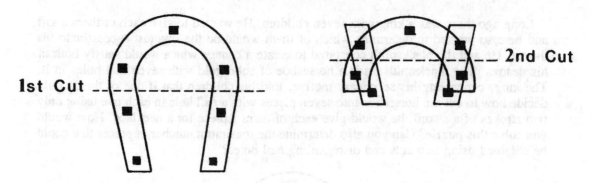

Notice: Piece #1 is on top of piece #3 in this diagram.

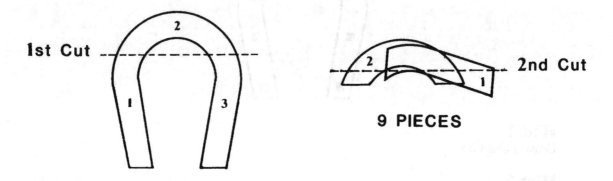

9 PIECES

39. QUICK CANDLESTICK

The Lopez family lives in an area which experiences violent thunderstorms that are frequently followed by blackouts or temporary losses of electricity. To prepare for such emergencies, the Lopez family keeps a supply of candles in two different sizes. Some of the candles are one inch longer than others.

During a blackout one winter evening, they light a long candle at 5:00 and a short one at 6:30. At 9:00 both candles are the same length. Eventually, the short candle burns out at 10:30 and the long one at 11:00.

What are the lengths of the candles in their supply?

Hint 1
Draw a diagram. Label the heights of the candles in the diagram at the various times mentioned in the problem.

Hint 2
Change the question to "Compare the times it takes the two candles to burn a given length."

Hint 3
What are the conditions that you use to determine the burning times of the candles? They are the same length at 9:00; then one burns out at 10:30 and the other at 11:00.

Hint 4
Change the information given. Suppose the short candle were the same length as the long candle originally. How long should it burn in order for it to be the same length as the other at 9:00?

ANSWER: 8 inches and 9 inches.

40. PARTLY POPULAR

In a recent election in the small town of Aberdeen, Brummett received 53 more votes than Moore, 89 more votes than Smith, and 114 more votes than Ross. Altogether the candidates received 4,108 votes. How many did each candidate receive?

Hint 1
Solve a simplified related problem. For example, suppose two candidates received 25 votes in an election and the winner received 5 more votes than the loser. How many votes did each candidate receive?

Hint 2
Solve a simplified related problem. Suppose that three candidates received twenty-five votes with the victor's tally exceeding that of the opponents by two and three votes.

Hint 3
Solve a simplified related problem. Suppose that four candidates received twenty-five votes with the victor's tally exceeding that of her opponents by one, two, and four votes.

ANSWER: Brummett received 1,091 votes, Moore received 1,038, Smith 1,022, and Ross 977 votes.

41. AN ODD GIFT?

At the North Pole Santa Claus has 1000 elves at work packing 1000 containers of gifts for little girls and boys. Since children were not equally nice, Santa plans to give different numbers of gifts depending upon how nice (or how naughty) each child was during the year. He instructs his elves to pack the containers in the following way: the first elf is to put a gift into each of the 1000 containers, the second elf is to put a gift into every second container, the third elf is to put one into every third container, and so on until all the elves have finished their tasks.

There are lots of different and interesting questions we might consider, but first let's look at this one: Which containers have an odd number of gifts? Of the children receiving the following containers, number 72, number 216, number 743, and number 784, which has been the nicest, second nicest, third nicest, and naughtiest?

Hint 1
Solve a simplified related problem. For example, apply the same conditions and ask the same question for a group of 10 or 20 elves.

Hint 2
Make a chart.

Hint 3
Search for a pattern using your chart.

Hint 4
Use all given and implied information. How can you determine the number of gifts held in any particular container?

ANSWER: The containers whose numbers are perfect squares.
The person who gets container #216 was nicest, followed by the ones who get #784, #72, and #743.

42. FAIR BALL?

St. Michael's Sluggers and Paul's Pizza Players of the slow pitch division of the County Amateur Softball League each have 26 wins, three losses, and one game left to play. Paul's Pizza Players can win the division championship during the regular season, but the Sluggers at best can finish in a tie for the first place. How can this be a fair solution?

RULES: (1) Individual games are played until one team wins; there are no ties.
(2) All 30 games each team is scheduled to play are played and count toward the championship.

Hint 1
Solve part of the problem by working backwards considering how each case can arise. (a) Paul's Pizza Players win, (b) St. Michael's Sluggers tie.

Hint 2
Introduce an element. For example, could other teams be involved?

ANSWER: A third team is involved in a three-way tie with St. Michael's and Paul's Pizza and is scheduled to play Paul's Pizza in the final game.

43. CREATING CRANAPPLE

Kirk was two large bottles, one with a gallon of apple juice, the other with a gallon of cranberry juice. Kirk measures one cup of apple juice and pours it into the apple juice. After stirring the apple-cranberry mixture, Kirk measures one cup of it and pours it into the apple juice.

Kirk wonders, "Is there more apple juice in the bottle that originally held just cranberry juice, or is there more cranberry juice in the apple juice bottle? Maybe there's the same amount of cranberry juice in the apple juice bottle as there is apple juice in the cranberry juice bottle; I'm just not sure."

Explain which case is correct?

Hint 1
Draw a diagram showing what happens at each step.

Hint 2
Solve part of the problem. After the cup of apple juice is poured into the cranberry juice, what is the ratio of apple juice to the liquid content of the cranberry juice bottle?

ANSWER: There is the same amount of cranberry juice in the apple juice bottle as apple juice in the cranberry juice bottle.

44. RELATIVELY UNKNOWN

John's dad enjoys number puzzles. One morning just before a visit from some distant relatives whom John had never met, his dad said, "My second cousin will be visiting on his fortieth birthday today. His three children's ages multiplied together equal his age and added together equal yours. Can you tell how old the children are?"

John enjoys his dad's number puzzles and usually solves them quickly, but this one stumped him. Then a car pulled into the driveway and two people got out. John's dad said, "The two older children are the first ones out."

John smiled and replied, "Now I can tell you their ages."

What were the ages of the children?

Hint 1
Make a chart of the possible ages of the children.

Hint 2
Use all given and implied information. For example, what does the statement, "John enjoys his dad's number puzzles and usually solves them quickly, but this one stumped him," imply about John's age?

Hint 3
What information is given in the problem? For example, what does, "The two older children are the first ones out," mean?

ANSWER: 8, 5, and 1.

45. NAME DROPPING

Roosevelt School, wishing to assess parents' attitudes on its curriculum innovations, decided to send a questionnaire to a random sampling of parents of its students. Stephen Culwell, who worked in the school office, was told to select about four percent of the parents to receive the questionnaires.

Culwell decided to utilize the alphabetical mailing list of parents in the following way. He would pick the first name on the list, skip one name, pick the next, skip two, pick the next, skip three, and so on.

Surprisingly enough, the last name on the list was the last one picked according to Culwell's scheme. Not only that, Culwell discovered that he had picked exactly four percent of the parents.

How many parents had Culwell picked?

Hint 1
Have you seen the same or a similar type of problem before, for example, Aw Sum Gauss?

Hint 2
Change the question. How many parents did Stephen skip?

Hint 3
Draw a chart listing the number picked, the number shipped, and the percent picked.

Hint 4
Search for a pattern in your chart.

ANSWER: Culwell picked 49 parents.

46. PARSIMONIOUS PURCHASER

Juanetta left home with a dollar bill given to her by her father. She planned to spend it all at a fair being held in her town. When she got home from the fair, she told her father she spent the whole dollar, for none of the purchases except the last did she have the exact change, each of the vendors made change for her so that she had the smallest possible number of coins after each purchase, and she made as many purchases as possible under these circumstances. Juanetta's father could not determine how many purchases his daughter made. Can you?

Hint 1
Solve a simplified related problem. Suppose Juanetta began with a dime. How many purchases could she make?

Hint 2
What are the conditions of the problem?

Hint 3
Use all given and implied information. What is the smallest coin that can be broken into smaller denominations?

Hint 4
Use the trial and error method.

ANSWER: 21 purchases (achieved in a variety of possible ways, but all based on the fact that the nickel is the smallest denomination which can be broken into smaller denominations).

47. INSTALLATION FREE

Susan's parents bought a country house and they had very few furnishings for it. One of their neighbors gave them a used rug that measured seven feet by six feet, but it had a hole which measured one foot by two feet in the middle. Susan said, "We can cut the rug into two pieces and sew them together to make it a perfect fit for our little eight foot by five foot room." How would she cut the rug?

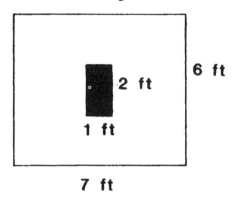

Hint 1
What are you asked to find?

Hint 2
Solve a simplified related problem. Suppose the rug is 3 feet by 3 feet and has a 1 foot by 2 foot hole in it. How can Susan cut the rug into two pieces that can be sewed together into a rug that measures 4 feet by 2 feet?

Hint 3
Solve a simplified related problem. Suppose Susan had a 5 x 4 foot rug with a 1 x 2 foot hole. How could she cut it into two pieces which she could sew into a 6 x 3 foot rug?

ANSWER:

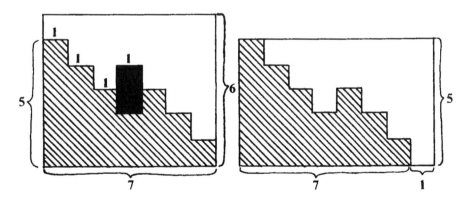

48. QUINTILLION PLUS

Pat, a girl who is fascinated by number patterns, thinks she has found an interesting sequence of numbers. She noticed that 49, 4489, 444889, ... can all be written as the product of a number times itself. What number multiplied by itself produces the tenth term in Pat's sequence?

Hint 1
Solve part of the problem. For example, how many digits are in the number which, when multiplied by itself, yields a product of 4489?

Hint 2
Use all given and implied information. For example, what should be the last digit of the number which, multiplied by itself, yields a product of 4489?

Hint 3
Search for a pattern in the numbers which are multiplied by themselves.

ANSWER: 6,666,666,667 multiplied by itself produces the tenth term.

49. THREE-QUARTER DIME JINGLE

When Danny came home after collecting money for UNICEF at Halloween his parents laughed. "You certainly do make a lot of noise as you walk!" his father said.

"I know," replied Danny. "I put the money I collected for UNICEF into my pockets and it jingles when I walk."

"Do you know how much you collected?" asked Danny's mother.

"I got a little more than twenty dollars, all in quarters and dimes. I put half the amount in each pocket. In my left pocket I've got the same number of dimes as quarters, but in the right pocket there's twice as much money in quarters as there is in dimes."

How many coins of each type did Danny have in each of his pockets?

Hint 1

Use all given and implied information. For example,

(a) What is the total amount collected?

(b) Given that he has the same amount in each pocket, what is implied about the total number of quarters and what is implied about the total amount collected?

(c) Given that he has the same number of quarters as dimes in the left pocket, what is implied about the amount of money in quarters and the amount in dimes in the left pocket?

(d) What is implied about the number of quarters and the number of dimes in the right pocket?

(e) What is implied about the amount in the right pocket and about the total amount in both pockets?

(f) What is implied now about the number of quarters in the left pocket and about the number of dimes in the left pocket?

(g) Now what can you say about the amount of money in the left pocket?

ANSWER: In the left pocket there are 30 dimes and 30 quarters. In the right pocket there are 35 dimes and 28 quarters.

50. NO HEADS IS BETTER THAN ONE

(a) Juan, Yolanda and Bob toss a coin 17, 18, and 19 times, respectively. Which one is least likely to have tossed more heads than tails?

(b) This time Juan, Yolanda, and Bob toss the coin 17, 19, and 20 times, respectively. Who is least likely to have tossed more heads than tails?

(c) Suppose the three friends toss the coin 18, 19, and 20 times, respectively. Who is least likely to have tossed more heads than tails?

Hint 1
Solve a simplified related problem. Suppose Juan, Yolanda, and Bob toss a coin one, two, and three times, respectively. Who is least likely to throw more heads than tails?

Hint 2
Solve part of the problem. Suppose Juan tosses a coin an odd number of times. What is the probability that he will toss more heads than tails?

Hint 3
Solve a simplified problem. Suppose Juan, Yolanda, and Bob toss a coin two, three, and four times, respectively. Who is least likely to toss more heads than tails?

Hint 4
If you plan did not lead to a solution, try a new plan. For example, change the question. If Yolanda tosses a coin an even number of times, say two or four times, what is the probability that she tosses an equal number of heads and tails? What is the probability that she tosses more tails than heads?

ANSWER:
(a) Yolanda
(b) Bob
(c) Juan